Part of the
NO-NONSENSE ALGEBRA Series

Algebra Word Problems Made Simple!

Master Algebra the Easy Way!

Richard W. Fisher

*America's Math Teacher, Richard W. Fisher,
will carefully guide you through each and every topic
with his award-winning system of teaching.*

Go to **www.NoNonsenseAlgebra.com**
for instant access to the Online Video Lessons.

Math
Essentials
MARINA, CALIFORNIA

Dedicated to Rachel . . . Physical Therapist Extraordinaire

Thank you for guiding me through my rehab.

Thanks even more for your kindness, encouragement, and inspiration.

First printing 2021
ISBN 978-0-9994433-4-7

Introduction

ABOUT THIS BOOK

What sets this book apart from other books is its approach. It is not just an algebra text, but a **system** of teaching algebra. Each of the short, concise, self-contained lessons contain four key parts:

1. A clear **introduction and explanation** of each new topic, written in a way that is easy for the student to understand

2. A **Helpful Hints** section that offers important tips and shortcuts

3. **Examples** with step-by-step solutions

4. **Written Exercises** with answers in a the back of the book

5. Each lesson contains plenty of **work space** on which to solve problems and take notes.

6. There is a **Solutions** box in which to place your answer.

Each topic contains the necessary structure and guidance that will ensure that the student will learn algebra in a systematic, step-by-step, logical manner. The book is set up in chapters, and there is a natural flow from each topic to the next. A comprehensive **Final Exam** is also included.

HOW TO USE THIS BOOK

Step 1

Carefully read the **Introduction** that begins each lesson. This section will include essential information about each new topic. Here you will find explanations as well as important terms and definitions.

Step 2

Carefully read the **Helpful Hints** section. This section will provide important tips and shortcuts. When completing the written exercises, it is often helpful to refer back to this section.

Step 3

Carefully go through the **Examples**. Each example shows the step-by-step process needed to complete each problem.

THE MOST IMPORTANT TIP THAT I CAN OFFER WHEN USING THIS BOOK!

COPY EACH EXAMPLE ON A PIECE OF PAPER. Than read and copy each of the steps. I promise that by doing this, you will find it much easier to understand the problem.

There is something very special about writing out an example and then copying out the steps. It makes the learning process so much more effective. When you do this, you are fully involved and will have a much deeper understanding. Just simply reading a problem and the steps is not nearly as effective.

Step 4

Work the written **Exercises**. If necessary, go back and re-read the **Introduction** and the **Helpful Hints**. You may want to go back and refer to the **Examples**, also.

Step 6

After completing the **Exercises**, correct your work. The **Solutions** section is located at the end of the book.

HOW TO USE THE FREE ONLINE VIDEO LESSONS

www.nononsensealgebra.com

Go to **www.nononsensealgebra.com** and click on VIDEO LIBRARY for full access to corresponding No-Nonsense Algebra video lessons, including Word Problems. The author, award-winning teacher, Richard W. Fisher, will carefully guide you through each topic, step-by-step. Each lesson will provide easy-to-understand instruction, and THE STUDENT CAN WORK EXAMPLES RIGHT ALONG WITH MR. FISHER. It's like having your own personal math tutor available 24/7. After the video lesson, you can go to the book and complete the lesson. The book combined with the video lessons will turbo-charge your ability to master algebra.

Table of Contents

Table of Contents

Keep These Tips in Mind When Solving Algebra Word Problems

Many students find that algebra word problems can be quite challenging. However, you will find that they are really quite easy. The secret is to solve them using an organized plan. There are several types of algebra word problems, and here are a few general tips to keep in mind when solving them.

- First, read the problem carefully and be sure that you fully understand it. Be sure you understand that which is given, and what is to be found.

- Second, select a variable to represent one of the unknowns. The variable will be used to describe all the other unknowns in the problem. Often it is good to have the variable represent the smallest number in the problem.

- Third, translate the word problem into an equation.

- Fourth, solve the equation and use the solution to answer the question that was asked for in the problem. Sometimes the answer will be the value of the variable. Sometimes it will be necessary to use the value of the variable to find what was asked for.

- Fifth, check your answers.

- You will find that sometimes using charts can be quite helpful when solving algebra word problems.

Chapter 1: Introduction to Algebra Word Problems

To solve **algebra word problems**, it is necessary to translate words into **algebraic equations** containing a **variable**. Keep in mind that a variable is a letter that represents a number.

In algebra word problems, the object is to find the **value** of the variable. To do this, you will put the algebraic skills that you have learned to use. Remember the following when translating words into algebraic expressions.

Helpful Hints:

- Some of the most common words indicating **addition**: sum, add, more, greater, increased, more than.

- Some of the most common words indicating **subtraction**: minus, less, less than, difference, decreased, diminished, reduced.

- Some of the most common words indicating **multiplication**: product, times, multiplied by.

- Some of the most common words indicating **division**: divide, quotient, ratio.

- **"Is"** often means "equal" (=).

Chapter 1: Introduction to Algebra Word Problems

To solve **algebra word problems**, it is necessary to translate words into **algebraic equations** containing a **variable**. A **variable** is a letter that represents a number. Here are some examples:

Three more than a number $\rightarrow$ **x + 3**

Four less than a number $\rightarrow$ **x – 4**

Twice a number $\rightarrow$ **2x**

Seven times a number $\rightarrow$ **7x**

The quotient of x and five $\rightarrow$ $\frac{x}{5}$

A number decreased by six $\rightarrow$ **x – 6**

Seven less than three times a number $\rightarrow$ **3x – 7**

One-third a number $\rightarrow$ $\frac{1}{3}$**x** or $\frac{x}{3}$

Twice a number less nine is equal to 15 $\rightarrow$ **2x – 9 = 15**

Two-fifths a number $\rightarrow$ $\frac{2}{5}$**x** or $\frac{2x}{5}$

The difference between three times a number and eight equals 12 $\rightarrow$ **3x – 8 = 12**

The sum of a number and -9 is 24 $\rightarrow$ **x + -9 = 24**

Three times a number less six equals twice the number plus 15 $\rightarrow$ **3x – 6 = 2x + 15**

Twice the sum of n and five $\rightarrow$ **2(n + 5)**

The difference between four times x and 15 equals twice the number $\rightarrow$ **4x - 15 = 2x**

Chapter 1: Introduction to Algebra Word Problems

Translate each of the following into an equation using the space provided. The workspace can also be used for notes.

1) Seven less than twice a number is 12.

   SOLUTION:

2) Two more than three times a number equals 30.

   SOLUTION:

3) The sum of twice a number and five is 14.

   SOLUTION:

4) The difference between four times a number and six is 10.

   SOLUTION:

Chapter 1: Introduction to Algebra Word Problems

5) Twelve is five less than four times a number.

SOLUTION:

6) One-third times a number less four equals twice the number added to eight.

SOLUTION:

7) Twice the sum of a number and two equals 10.

SOLUTION:

8) The difference between five times a number and three is 17.

SOLUTION:

EXERCISES

9) Twice a number decreased by six is 15.

    SOLUTION:

10) Two less than three times a number is seven more than twice the number.

    SOLUTION:

11) Four more than a number equals the sum of seven and -12.

    SOLUTION:

EXERCISES

12) A number divided by five is 25.

SOLUTION:

13) Twenty-five is nine more than four times a number.

SOLUTION:

14) Sixteen subtracted from five times a number is equal to five plus the number.

SOLUTION:

EXERCISES

15) The sum of four x and three is the same as the difference of two x and two.

SOLUTION:

16) Twice the quantity of two y and 7 equals 10.

SOLUTION:

17) Four times the quantity of x plus 9 is the same as the difference of x and eight.

SOLUTION:

EXERCISES

18) Fourteen equals the sum of six and a number divided by three.

SOLUTION:

19) If the product of five and a number is divided by 8, the result is 12.

SOLUTION:

20) Four times a number less 7 is equal to 12 more than two times that number.

SOLUTION:

Chapter 2: Everyday Algebra Word Problems

To solve an everyday **algebra word problem**, the basic idea is to **translate** the words into an **equation**. Once you have the equation, all you need to do is solve it and then check your answer. Use the following steps when solving everyday algebra word problems.

Helpful Hints:

- **First**, read the problem **carefully** and be sure that you fully understand it. Be sure you understand that which is **given**, and what is to be **found**.

- **Second**, select a variable to represent one of the unknowns. This variable will be used to describe all the other numbers in the problem. Often it is good to have the variable represent the smallest number in the problem.

- **Third**, translate the problem into an equation.

- **Fourth**, solve the equation and use the solution to answer the question that was asked in the problem. Sometimes the answer will be the value of the variable. Sometimes it will be necessary to use the value of the variable to find what was asked for in the problem.

- **Fifth**, check your answer.

Notes:

_____

_____

_____

_____

_____

Chapter 2: Everyday Algebra Word Problems

Solve each of the algebra word problems using a variable and an equation.

1) **The difference between three times a number and 9 is 15. Find the number.**

Select the variable   let x = the number
Write the equation   3x – 9 = 15
Solve the equation   3x – 9 + 9 = 15 + 9
                      3x = 24
                      x = 8

The number is 8. *Check your answer.*

2) **Four times a number less six is eight more than two time the number. Find the number.**

Select the variable   let x = the number
Write the equation   4x – 6 = 2x + 8
Solve the equation   4x – 2x – 6 = 2x – 2x + 8
                      2x – 6 + 6 = 8 + 6
                      2x = 14
                      x = 7

The number is 7. *Check your answer.*

3) **A board 44 cm long is cut into two pieces. The long piece is three times the length of the short piece. What is the length of each piece?**

Select the variable   let x = the short piece
                      3x = the long piece
Write the equation   3x + x = 44
Solve the equation   4x = 44
                      x = 11

The short piece, x, is 11 cm.
The long piece, 3x, is 33 cm. *Check your answers.*

4) **Roy weighs 50 kg more than Bill. Their combined weight is 170 kg. What is each of their weights?**

Select the variable   let x = Bill's weight
                      x + 50 = Roy's weight
Write the equation   x + (x + 50) = 170
Solve the equation   2x + 50 – 50 = 170 – 50
                      2x = 120
                      x = 60

Bill's weight, x, is 60 kg.
Roy's weight, x + 50, is 110 kg. *Check your answers.*

Chapter 2: Everyday Algebra Word Problems

	EXAMPLES

5) Find two consecutive integers whose sum is 91.

Select the variables   let x = first integer

$(x + 1)$ = second integer

Write the equation $x + (x + 1) = 91$

Solve the equation $2x + 1 = 91$

$2x + 1 - 1 = 91 - 1$

$2x = 90$

$x = 45$

The first integer, x, is 45.
The second integer, $x + 1$, is 46.
Check your answers.

6) Find three consecutive even Integers whose sum is 156.

Select the variable   let x = first integer

$(x + 2)$ = second integer

$(x + 4)$ = third integer

Write the equation $x + (x + 2) + (x + 4) = 156$

Solve the equation $3x + 6 = 156$

$3x + 6 - 6 = 156 - 6$

$3x = 150$

$x = 50$

The integers are
  $x = 50$
$x + 2 = 52$
$x + 4 = 54$ *Check your answers.*

Chapter 2: Everyday Algebra Word Problems

Solve each algebra word problem using a variable and an equation. Show all steps and write the solution in the box.

1) Six times a number less 7 is 41. Find the number.

   SOLUTION:

2) Eight more than three times a number is 194. Find the number.

   SOLUTION:

3) The difference between seven times a number and three times that same number is 20. Find the number.

   SOLUTION:

EXERCISES

4) Two more than three times a number is eight more than that number. Find the number.

SOLUTION:

5) One-third a number less three is 12. Find the number.

SOLUTION:

6) One number is twice the value of another number. Their sum is 96. Find the numbers.

SOLUTION:

Chapter 2: Everyday Algebra Word Problems

EXERCISES

7) Julie and John earned a total of 72 dollars. If Julie earned three times as much as John, how much did each of them earn?

SOLUTION:

8) In a school, the number of seventh graders is 215 more than the number of eighth graders. If the total of the two grades is 895 students, how many seventh graders and how many eighth graders are there?

SOLUTION:

9) Steve and Stan sold a total of 93 candy bars for a fund-raiser. If Steve sold 5 more than 3 times the number that Stan sold, find the number of candy bars sold by each.

SOLUTION:

EXERCISES

10) Find two consecutive integers whose sum is 125.

SOLUTION:

11) Find 3 consecutive integers whose sum is 99.

SOLUTION:

12) Find 3 consecutive odd integers whose sum is 159.

SOLUTION:

EXERCISES

13) Five times a number is twelve more than twice that number. Find the number.

SOLUTION:

14) The perimeter of a triangle is 30 feet. The lengths of its side are three consecutive integers. Find the lengths.

SOLUTION:

15) The larger of two numbers is 4 times the the smaller. Their sum is 105. Fine the numbers.

SOLUTION:

EXERCISES

16) A rectangle is twice as long as it is wide, and its perimeter is 72 feet. Find the length and the width.

SOLUTION:

17) Find two number. The second is 5 less than 3 times the first, and their sum is 31.

SOLUTION:

18) The sum of 3 consecutive odd integers whose sum is 375. Find the odd integers.

SOLUTION:

EXERCISES

19) The perimeter of a rectangle is 30 feet. If the length is 2x - 3, and the width is x. Find the length of each side.

SOLUTION:

20) A rectangular field is three times as long as it is wide. The perimeter is 392 feet. Find its length and width.

SOLUTION:

21) Five times Bob's age plus six, equals three times his age plus 30. Find Bob's age.

SOLUTION:

Chapter 3: Time, Rate, and Distance Problems

:---:	
INTRODUCTION	

Solving **time, rate, and distance problems** are easy if you approach them in an organized fashion. You need to know the following formula.

Distance = Rate x Time which is commonly written **D = R x T**

Also, you need to know the two related formulas: $T = \dfrac{D}{R}$ and $R = \dfrac{D}{T}$

Remember the following when solving time, rate, and distance problems.

Helpful Hints:

- **First**, read the problem **carefully** and be sure that you fully understand it. Be sure you understand that which is **given**, and what is to be **found**.

- Always draw a sketch. This makes it easier to understand the problem.

- Keep in mind the formula **Distance = Rate x Time**

- Use a chart or table to do your work. Assign a variable to one of the unknowns.

- There are **four** basic types of motion problems
  1. **Separation** problems where two objects start from the same place moving in opposite directions.
  2. **Come-together** problems where two objects start from different places and move towards each other.
  3. **Catch-up** problems where one object leaves a place and a second object leaves the same location at a later time and catches up.
  4. **Back-and-Forth** problems where an object goes out from a place, turns around and comes back using the same route.

- Check your answers.

Chapter 3: Time, Rate, and Distance Problems

Solve each of the following. Identify the type, draw a sketch, and use a chart.

1) **Two cars are 360 km apart. They travel toward each other, one at 80 km per hour and the other at 100 km per hour. How much time will it take before they meet?**

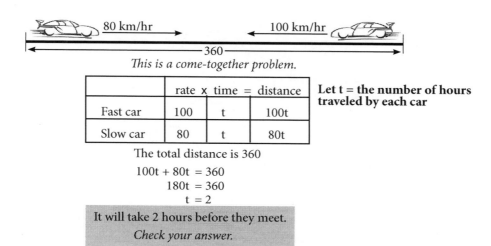

This is a come-together problem.

	rate	x time	= distance
Fast car	100	t	100t
Slow car	80	t	80t

Let t = the number of hours traveled by each car

The total distance is 360

$$100t + 80t = 360$$
$$180t = 360$$
$$t = 2$$

It will take 2 hours before they meet.
Check your answer.

2) **Bill and Annie each leave home driving in opposite directions for 3 hours and are then 510 km apart. If Bill's speed was 80 km per hour, what was Annie's speed?**

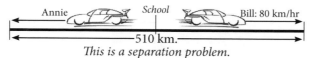

This is a separation problem.

	rate	x time	= distance
Bill	80	3	240
Annie	r	3	3r

Let r = Annie's rate

The total distance is 510 km.

$$3r + 240 = 510$$
$$3r + 240 - 240 = 510 - 240$$
$$3r = 270$$
$$r = 90$$

Annie's speed (rate) was 90 km/hr.
Check your answer.

Chapter 3: Time, Rate, and Distance Problems

EXAMPLES

3) Mary left her house by bicycle, traveling 40 km per hour. Two hours later, John left the same house in a car trying to catch up with Mary. If he was traveling at a rate of 60 km per hour, how long would it take John to catch up with Mary?

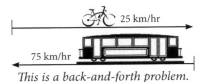
House — Mary: 40 km/hr — John: 60 km/hr

This is a catch-up problem.

	rate	x time	= distance
Mary	40	t+2	40 (t + 2)
John	60	t	60t

Let t = the number of hours that John will travel

They each traveled the same distance.

$$60t = 40 (t + 2)$$
$$60t = 40t + 80$$
$$60t - 40t = 40t - 40t + 80$$
$$20t = 80$$
$$t = 4$$

John caught up in 4 hours.
Check your answer.

4) Ron rides his bike from home to the lake traveling at a rate of 25 km per hour. He returns home by train at a rate of 75 km per hour. If the bike trip took 2 hours longer than the train trip, how far was it from his home to the lake?

25 km/hr
75 km/hr

This is a back-and-forth problem.

	rate	x time	= distance
Bike	25	t+2	25 (t + 2)
Train	75	t	75t

Let t = the number of hours traveled by the train

The distances are equal.

$$75t = 25 (t + 2)$$
$$75t = 25t + 50$$
$$75t - 25t = 25t - 25t + 50$$
$$50t = 50$$
$$t = 1$$

Check: 75t = 75(1) = 75
25 (t + 2) = 25(3) = 75

The distance from home to the lake is 75 km.
Check your answer.

Chapter 3: Time, Rate, and Distance Problems

EXAMPLES

5) **A car and a motorcycle start traveling towards each other at the same time from locations 405 km apart. The rate of the car is twice the rate of the motorcycle. In 3 hours they pass each other. Find the rate of each.**

This is a come-together problem.

	rate	x	time =	distance
Motorcycle	r	3		3r
Car	2r	3		6r

The total distance is 405 km.

$$3r + 6r = 405$$
$$9r = 405$$
$$r = 45$$

The rate of the motorcycle is r = 45 km/hr.
The rate of the car is 2r = 90 km/hr.

Check your answers.

Notes:

_____

_____

_____

_____

_____

_____

_____

_____

_____

_____

Chapter 3: Time, Rate, and Distance Problems

1) Ralph and Joe leave school walking in opposite directions. Ralph walks 1 km per hour faster than Joe. After 2 hours they are 30 km apart. What was the rate of each?

SOLUTION:

	rate x time = distance		

2) Two trains left a station traveling in opposite directions. One traveled at the rate of 60 km per hour, and the other at 72 km per hour. How many hours passed before they were 792 km apart?

SOLUTION:

	rate x time = distance		

3) Two trucks are 180 km apart. They each left at the same time traveling towards each other. One traveled at a rate of 65 km per hour, and the other at 55 km per hour. How many km did each travel before they met?

SOLUTION:

	rate	x	time	=	distance

4) Two planes left the same airport, traveling in opposite directions. One plane traveled 60 km per hour faster than the other. After 5 hours they were 5300 km apart. Find the rate of each.

SOLUTION:

	rate	x	time	=	distance

5) The first runner started a race and maintained a rate of 20 km per hour. One hour later a second runner started the race and maintained a rate of 25 km per hour. How many hours passed before the second runner caught up with the first runner?

SOLUTION:

	rate x time = distance		

6) Steve walked from his house to the lake at a rate of 6 km per hour. He rode a bike back to his house at the rate of 18 km per hour. The walk took 4 hours longer than the bike ride. How far is it from his home to the lake?

SOLUTION:

	rate x time = distance		

EXERCISES

7) Elena left her house driving her car at the rate of 45 km per hour. Two hours later, her sister Yana left the house traveling the same direction at a rate of 60 km per hour. How many hours will it take for Yana to catch up with Elena?

SOLUTION:

	rate	x	time	=	distance

8) Sally spent 6 hours walking from her house to the lake and back. She walked to the lake at a rate of 4 km per hour, and walked back to her house at the rate of 2 km per hour. What is the distance from her house to the lake?

SOLUTION:

	rate	x	time	=	distance

Chapter 3: Time, Rate, and Distance Problems

9) Bob rode his bike south at the constant rate of 28 mph. Susan started from the same location and rode her bike north at the constant rate of 22 mph. If they ride at a constant rate, in how many hours will they be 125 miles apart?

SOLUTION:

	rate	x	time	=	distance

10) Mr. Allen and Mrs. Benson start at the same time from two locations 288 miles apart. They drive towards each other. Mr. Allen travels eight miles an hour faster than Mrs. Benson. At the end of six hours they meet. What is the rate of each?

SOLUTION:

	rate	x	time	=	distance

What About the Problems that I Get Wrong?

When learning algebra, every student makes mistakes.

It is important to find out why you worked a problem incorrectly, so that you don't make the same mistake in the future.

Here are some tips to help you correct your mistakes.

■ Many mistakes are based on careless errors. Carefully re-working the problem can usually result in finding these errors.

Here are just a few common careless errors:

- The student did not carefully read the problem and understand what was asked.

- The problem was not copied or set up correctly.

- There were mistakes with positive or negative signs.

- There were mistakes in the order of operations.

- Fractions were not reduced to lowest terms.

- Answers were not in their simplest form.

- Careless computational errors were made.

■ Sometimes the student does not fully understand the topic. This will result in difficulties with the problems. If this is the case, it would be good to consider doing any or all of the following:

- View and the Online Tutorial Video again, being careful to work along with the instructor.

- Re-read the Introduction and Helpful Hints section.

- Study the Examples pertaining to the topic, being careful to copy the problem as well as all the steps.

Chapter 4: Mixture Problems

There are many everyday problems that involve the mixing of products that have different costs. **Mixture problems** have many applications in business, industry, science, and even supermarkets. When working with mixtures it is important to keep in mind the following formula. **Number of items x Price per item = Cost**

It is also very helpful to make a chart when solving mixture problems. Keep the following in mind when solving mixture problems.

Helpful Hints:

- **First**, read the problem **carefully** and be sure that you fully understand it. Be sure you understand that which is **given**, and what is to be **found**.

- Use a **chart**.

- Assign a **variable** to one of the unknowns.

- It if often helpful to express some costs as **cents**. Example: $2.75 is 275 cents. It helps avoid having to work with decimals.

- Some mixture problems involve **percents**. It might be good to review the lesson on percents in chapter 1.

- It is often good to use a **calculator** to do the computation.

- Check your answers.

Chapter 4: Mixture Problems

EXAMPLES

Use a chart to solve each of the mixture problems.

1) **A storeowner wants to mix one type of candy worth $2.60 per kg with another type of candy worth $3.60 per kg. He wants 40 kg of a mixture that is worth $3.00 per kg. How many kg of each should he use?**

	# of kg	×	price per kg	=	cost
$3.60 Candy	n		360		360n
$2.60 candy	40 – n		260		260(40 – n)
Mixture	40		300		40 (300)

Let n = the number of kg of the $3.60 candy

Notice that we changed the dollars to cents

The value of the mixture is equal to the value of the $3.60 candy added to the value of the $2.60 candy. Notice the values are changed to cents.

$$360n + 260 (40 - n) = 40(300) \qquad \textit{Use a calculator.}$$

$$360n + 10{,}400 - 260n = 12{,}000$$

$$360n + 10{,}400 - 10{,}400 - 260n = 12{,}000 - 10{,}400 \qquad \textit{Subtract 10,400 from each side.}$$

$$360n - 260n = 1{,}600$$

$$100n = 1{,}600$$

$$n = 16$$

$$(40 - n) = 24$$

> 16 kg of $3.60 candy
> 24 kg of $2.60 candy
> *Check your answers.*

Notes:

_____

_____

_____

_____

_____

_____

_____

_____

Chapter 4: Mixture Problems

2) **A scientist has a solution which is 15% pure acid. He has a second solution which is 40% pure acid. How many liters of each solution should he use to make 1200 liters of a solution which is 25% pure acid?**

	# of liters x	% of acid =	amount of pure acid
First solution	n	.15	.15n
Second solution	1200 – n	.4	.4(1200 – n)
Mixture	1200	.25	.25(1200)

Let n = the number of liters of the 15% solution

The sum of the amount of the solutions will equal the amount of the mixture

$$.15n + .4(1200 - n) = .25(1200)$$ *Use a calculator.*

$$.15n + 480 - .4n = 300$$

$$.15n - .4n + 480 - 480 = 300 - 480$$ *Subtract 480 from each side.*

$$-.25n = -180$$

$$.25n = 180$$

$$n = 720$$

n = 720 liters of the first solution

(1200 – n) = 480 liters of the second solution

Check your answers.

Notes:

_____

_____

_____

_____

_____

_____

_____

_____

_____

Chapter 4: Mixture Problems

EXERCISES

Use a chart to solve each of the following mixture problems.

1) A company mixes two kinds of cleansers to get a blend selling for 59 cents per liter. One kind is 50 cents per liter, and the other is 80 cents per liter. How much of each should be mixed to get 1,000 liters of blend?

SOLUTION:

2) Two kinds of candy are mixed to sell at $4.00 per kg. Vanilla candy sells for $3.20 per kg, and chocolate candy sells for $4.40 per kg. How much of each kind is used to make a mixture that weighs 60 kg?

SOLUTION:

3) A restaurant has a soup which is 24% cream and another soup which is 18% cream. How many liters of each must be used to make 90 liters of soup which is 22% cream?

SOLUTION:

4) A chemist has one solution which is 30% pure acid and another solution which is 60% pure acid. How many liters of each solution must be used to make 60 liters of solution which is 50% pure acid?

SOLUTION:

EXERCISES

5) How many pounds of peanuts worth $.90 per pound must be added to 10 pounds of walnuts worth $.60 a pound to form a mixture worth $.80 per pound?

SOLUTION:

6) How many ounces of a 75% acid solution must be added to 30 once of a 15% acid solution to make a 50% acid solution?

SOLUTION:

Chapter 4: Mixture Problems

EXERCISES

7) Tickets for a play are $30 for adults and $25 for children. If 52 people attended for a total cost of $1360, how many of the attendees were adults?

SOLUTION:

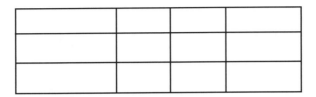

8) How many ounces of a 30% solution of acid must be added to 40 ounces of a 12% solution to produce a 20% solution?

SOLUTION:

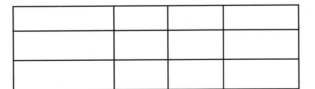

Some Easy Ways to Stay Tuned-Up!

You are learning lots of algebra topics in this book. As you work through each chapter, you might find it helpful to go back occasionally and do some review on your own. Doing this can keep the topics that you have studied, fresh in your mind.

Here are a few tips that can help you to remember what you have learned.

- Watch the Online Video Tutorials a second time. Don't be passive. Work right along with the instructor.

- For each lesson that you have completed, go back and review the Introduction, Helpful Hints, and Examples.

- Another helpful learning tool is the Glossary, which is located in the back of the book. The glossary contains important terms and definitions. It is a good place to review topics that you have learned, and also it can give you a brief introduction to terms that you will be studying later.

Chapter 5: Work Problems

INTRODUCTION

Work problems involve the time it takes for people or machines to complete jobs. These problems can apply to such jobs as painting, completing office tasks, construction, manufacturing, and much more. It is good to use a **chart** when completing work problems. Remember the following.

Helpful Hints:

- **Rate of work × Time = Work done**

- **First**, read the problem **carefully** and be sure that you fully understand it. Be sure you understand that which is **given**, and what is to be **found**.

- Use the same units of measure. For example, don't mix days with minutes.

- Fractions are used to show parts of the job.

- Be sure to understand what is being asked for, and answer that question.

- Check your answers.

Notes:

_____

_____

_____

_____

_____

_____

_____

_____

Chapter 5: Work Problems

Solve each of the following.

1) Mary can clean the house in 2 hours. Tom can clean it in 3 hours. If the work together, how long will it take to clean the house?

	Mary	Tom	Together
Hours needed	2	3	n
Part completed in one hour	$\frac{1}{2}$	$\frac{1}{3}$	$\frac{1}{n}$

Part Mary does in one hour	+	Part Tom does in one hour	=	Part of job done in one hour
$\frac{1}{2}$	+	$\frac{1}{3}$	=	$\frac{1}{n}$

Let n = the number of hours needed to complete the job when they work together

$$\frac{1}{2} + \frac{1}{3} = \frac{1}{n}$$

$$6n \cdot \frac{1}{2} + 6n \cdot \frac{1}{3} = 6n \cdot \frac{1}{n} \qquad \textit{Multiply each term by LCM = 6n}$$

$$3n + 2n = 6 \qquad \textit{Solve for n}$$

$$5n = \frac{6}{5}$$

$$n = 1\frac{1}{5}$$

> From the chart, n = the number of hours to complete the job together = $1\frac{1}{5}$ hours.

2) John can paint a room in 5 hours. If he and Ellen together can paint it in 2 hours, how long would it take Ellen to paint the room alone?

	John	Ellen	Together
Hours needed	5	x	2
Part completed in one hour	$\frac{1}{5}$	$\frac{1}{x}$	$\frac{1}{2}$

Part John does in one hour	+	Part Ellen does in one hour	=	Part of job done in one hour
$\frac{1}{5}$	+	$\frac{1}{x}$	=	$\frac{1}{2}$

Let x = the number of hours needed by Ellen to complete the job

$$\frac{1}{5} + \frac{1}{x} = \frac{1}{2}$$

$$10x \cdot \frac{1}{5} + 10x \cdot \frac{1}{x} = 10x \cdot \frac{1}{2} \qquad \textit{Multiply each term by LCM = 10x}$$

$$2x + 10 = 5x$$

$$5x = 2x + 10 \qquad \textit{Solve for x}$$

$$5x - 2x = 2x - 2x + 10$$

$$3x = 10$$

$$x = 3\frac{1}{3}$$

> From the chart, x = the number of hours Ellen needs to complete the job alone = $3\frac{1}{3}$ hours.

EXAMPLES

3) A pump can fill a tank in 5 hours. Another pump can empty the tank in 6 hours. If both pumps are on how many hours are needed to fill the tank?

Let x = the number of hours needed to fill the tank with both pumps open.

	Time	Part	Part in hours
Fill	5	$\frac{1}{5}$	$\frac{x}{5}$
Empty	6	$\frac{1}{6}$	$\frac{x}{6}$

$$\begin{array}{ccccc} \text{Amount} & & \text{Amount} & & \\ \text{Filled} & - & \text{Emptied} & = & \text{Full Pool} \\ \frac{x}{5} & & \frac{1}{6} & & 1 \end{array}$$

Keep in mind that each pump works for x hours. The tank fills faster than it empties.

$$30 \cdot \frac{x}{5} - 30 \cdot \frac{x}{6} = 30 \cdot 1 \quad \textit{Multiply by LCM = 30}$$
$$6x - 5x = 30$$
$$x = 30$$

It would take 30 hours to fill the tank.

Notes:

_____

_____

_____

_____

_____

_____

_____

_____

_____

_____

Chapter 5: Work Problems

Solve each of the following. Use a chart.

1) Julie can clean the windows of a house in 3 hours. Her brother, John, can clean the windows in 6 hours. How long will it take if they work together?

SOLUTION:

2) Steve can mow a lawn in 30 minutes. Jose can mow the same lawn in 20 minutes. How long will it take if they work together?

SOLUTION:

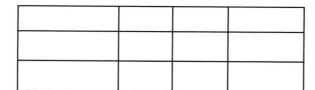

EXERCISES

3) A father and his son can paint a house in 3 days. Working alone, the father can paint the house in 4 days. How long would it take the son if he worked alone?

SOLUTION:

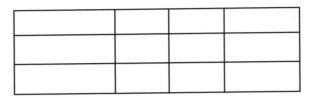

4) A pipe can fill a swimming pool in 3 hours. Another pipe can empty the swimming pool in 6 hours. If the pool is empty and both pipes are opened, how long will it take to fill the swimming pool?

SOLUTION:

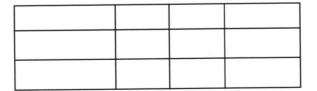

Chapter 5: Work Problems

EXERCISES

5) Susan can paint a room in 6 hours. If her sister helps her, the job is finished in 4 hours. How long would it take Susan's sister to paint the room if she worked alone?

SOLUTION:

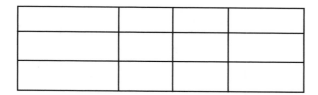

6) Dan and Dave can work together and paint a house in 4 days. Dan works twice as fast as Dave. How long would it take each of them to paint the house alone?

SOLUTION:

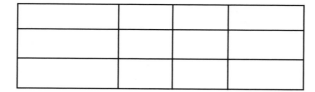

Chapter 5: Work Problems

7) Phil can build a fence in 2 hours. Steve can build the fence in 6 hours. Phil painted alone for 1 hour and had to quit working. How long would it take Steve to complete the job?

SOLUTION:

8) One secretary can type a report in 12 hours. Another secretary can do the job in 18 hours. How long would it take if both secretaries worked together to complete the job?

SOLUTION:

Algebra Word Problems Made Simple!

One of the Most Important Tips of All!

This tip applies to this book, and to any other math book
that you use in the future.

Here it is.

Most math books contain examples that show the necessary steps
in solving a problem. Many students merely read through the
examples. Then, when they have finished reading and attempt to do
the written exercises, they experience difficulties, and wonder why.

This is what you need to do. First, neatly copy the example on a
sheet of paper. Next, also copy each and every step.

I promise that by doing this, you will find it much easier to
understand the problem, and you will remember the process
necessary to solve it.

There is something very special about writing a problem down and
then writing out the steps. It makes the learning process so much
more effective. When you do this, you are fully involved and will
experience a much deeper understanding.

Just simply reading a problem and the steps is not nearly as
effective.

This is a Very Simple Tip. But It Works!

Chapter 6: Age Problems

Age problems can be solved with or without a chart. We will approach the age problems without using a chart. You will have to represent **past** and **future** ages when working these problems. To represent a **past** age, simply **subtract** from the present age. To represent a **future** age, simply **add** to the present age. Use the following steps when solving age problems.

Helpful Hints:

- **First**, read the problem **carefully** and be sure that you fully understand it. Be sure you understand that which is **given**, and what is to be **found**.

- **Second**, select a variable to represent one of the unknowns. This variable will be used to describe all the other numbers in the problem. Often it is good to have the variable represent the smallest number in the problem.

- **Third**, translate the problem into an equation.

- **Fourth**, solve the equation and use the solution to answer the question that was asked in the problem. Sometimes the answer will be the value of the variable. Sometimes it will be necessary to use the value of the variable to find what was asked for in the problem.

- **Fifth**, check your answers.

Chapter 6: Age Problems

Solve each of the following.

1) **Steve is 6 years older than Al. If the sum of their ages is 30, find each of their ages.**

Let x = Al's age
$x + 6$ = Steve's age

$x + (x + 6) = 30$ *Write the equation and solve.*

$2x + 6 = 30$

$2x + 6 - 6 = 30 - 6$

$2x = 24$

$x = 12$

> Al's age = x = 12
> Steve's age = $x + 6$ = 18
> *Check your answers.*

2) **Angela is 3 times as old as Linda. In 5 years Angela will be twice as old as Linda will be. Find their present ages.**

Let x = Linda's age $(x + 5)$ = Linda's age in 5 years
Let $3x$ = Angela's age $(3x + 5)$ = Angela's age in 5 years

$3x + 5 = 2(x + 5)$ *Write the equation and solve.*

$3x + 5 = 2x + 10$

$3x - 2x + 5 = 2x - 2x + 10$

$x + 5 = 10$

$x + 5 - 5 = 10 - 5$

$x = 5$

> Linda's age = x = 5
> Angela's age = $3x$ = 15
> *Check your answers.*

Chapter 6: Age Problems

3) **Juan is 5 years older than Amir. Five years ago, Juan was twice as old as Amir. Find their present ages.**

Let x = Amir's age
Let x + 5 = Juan's age

$$(x + 5) - 5 = 2 (x - 5)$$ *Write the equation and solve.*
$$x = 2x - 10$$
$$x - x = 2x - x - 10$$
$$0 = x - 10$$
$$10 = x$$

Amir's age = x = 10
Juan's age = x + 5 = 15
Check your answers.

4) **Elena's mother is 26 year older than Elena. In 10 years the sum of their ages will be 80. What are their ages now?**

Let x = Elena's age
x + 26 = Mother's age

$$(x + 10) + (x + 26 + 10) = 80$$ *Write the equation and solve.*
$$2x + 46 = 80$$
$$2x + 46 - 46 = 80 - 46$$
$$2x = 34$$
$$x = 17$$

Elena's age = x = 17
Mother's age = x + 26 = 43
Check your answers.

Chapter 6: Age Problems

Solve each of the following.

1) A mother is 6 times as old as her son. In 6 years, the mother will be 3 times as old as her son. What are their present ages?

> SOLUTION:

2) Moe is 8 years older than Zach. If twenty years ago Moe was three times as old as Zach, what are their present ages?

> SOLUTION:

Chapter 6: Age Problems

3) Dave is now twice as old as Chuck. Six years ago, Dave was three times as old as Chuck. Find their present ages.

SOLUTION:

4) Laura is 25 years older than Larry. In 10 years from now, Laura will be twice as old as Larry will be. Find their present ages.

SOLUTION:

Chapter 6: Age Problems

EXERCISES

5) Sam is now 7 years older than Lance. In 20 years, the sum of their ages will be 81. What are their present ages?

SOLUTION:

6) Vica is twice as old as Carly. Ten years ago, the sum of their ages was 70. How old is each of them now?

SOLUTION:

EXERCISES

7) Sophie is 4 times as old as Rhoda. In three years, Sophie will be three times as old as Rhoda. What are their present ages?

SOLUTION:

8) Eric is now four times as old as Ralph. Five years ago, Eric was nine times as old as Ralph. How old is each of them now?

SOLUTION:

Chapter 6: Age Problems

EXERCISES

9) A father is 40 years older than his son. In 10 years, the sum of their ages will be 80. What are their present ages?

SOLUTION:

10) Jake is three times as old as Olivia. Four years ago, Jake was 4 times as old as Olivia was at that time. What are their present ages?

SOLUTION:

Algebra Word Problems Made Simple!

61

Chapter 7: Coin Problems

When working with **coin problems** it is important to keep the **value** of the coins in mind. For example, three nickels has a value of 3(5) = fifteen cents. When solving coin problems it is usually a good idea to express the values in **cents**. For example, $3.25 would be expressed as 325 cents. Remember the following when solving coin problems.

Helpful Hints:

- **First**, read the problem **carefully** and be sure that you fully understand it. Be sure you understand that which is **given**, and what is to be **found**.

- **Second**, select a variable to represent one of the unknowns. This variable will be used to describe all the other numbers in the problem. Often it is good to have the variable represent the smallest number in the problem.

- **Third**, translate the problem into an equation. Remember to keep the **value** of the coins in mind when writing the equation.

- **Fourth**, solve the equation and use the solution to answer the question that was asked in the problem. Sometimes the answer will be the value of the variable. Sometimes it will be necessary to use the value of the variable to find what was asked for in the problem.

- **Fifth**, check your answers.

Chapter 7: Coin Problems

Solve each of the following.

1) A man has twice as many quarters as nickels. The total value of the coins is $4.40. How many of each coin does he have?

Let n = the number of nickels
2n = the number of quarters

$$5n + 25(2n) = 440$$ *Write the equation and solve.*
$$5n + 50n = 440$$ *Keep the value of each coin in mind*
$$55n = 440$$
$$n = 8$$

> n = 8 nickels
> 2n = 16 quarters
> *Check your answers.*

2) A boy has saved 58 coins consisting of dimes and nickels. The total value of the coins is $4.80. How many of each coin has he saved?

Let n = the number of nickels
58 − n = the number of dimes

$$5n + 10(58 - n) = 480$$ *Write the equation and solve.*
$$5n + 580 - 10n = 480$$ *Keep the value of each coin in mind*
$$-5n + 580 = 480$$
$$-5n + 580 - 580 = 480 - 580$$
$$-5n = -100$$
$$\frac{-5n}{-5} = \frac{-100}{-5}$$
$$n = 20$$

> n = 20 nickels
> 58 − n = 38 dimes
> *Check your answers.*

EXAMPLES

3) **A girl has a collection of nickels, dimes, and quarters whose total value is $11.25. She has 3 times as many nickels as dimes, and 5 more quarters than dimes. How many of each coin are in her collection?**

Let x = the number of dimes
3x = the number of nickels
x + 5 = the number of quarters

$$10x + 5(3x) + 25(x + 5) = 1125$$ *Write the equation and solve.*

$$10x + 15x + 25x + 125 = 1125$$ *Keep the value of each coin in mind*

$$50x + 125 = 1125$$

$$50x + 125 - 125 = 1125 - 125$$

$$50x = 1000$$

$$x = 20$$

x = 20 dimes
3x = 60 nickels
x + 5 = 25 quarters
Check your answers.

Notes:

_____

_____

_____

_____

_____

_____

_____

_____

_____

Chapter 7: Coin Problems

EXERCISES

Solve each of the following.

1) A girl has a collection of dimes and quarters. She has 4 times as many quarters as dimes. If the total value of her collection is $2.20, how many dimes and how many quarters does she have?

SOLUTION:

2) A man has 60 coins made up of dimes and quarters. If the total value is $9.60, how many of each of the coins does he have?

SOLUTION:

Chapter 7: Coin Problems

3) A woman has 7 times as many dimes as nickels. The total value of the coins is $3.75. How many of each coin does she have?

SOLUTION:

4) A boy has 3 times as many dimes as nickels. The total value is $2.80. How many of each coin does he have?

SOLUTION:

EXERCISES

5) A collection of 40 dimes and quarters has a value of $6.40. How many of each coin type are in the collection?

SOLUTION:

6) A man has a collection of nickels, dimes, and quarters whose value is $6.55. He has 3 times as many quarters as dimes, and 5 more nickels than dimes. How many of each type of coin does he have?

SOLUTION:

Chapter 7: Coin Problems

EXERCISES

7) A woman's purse contains nickels and dimes whose total value is $2.70. If there are 30 coins in all, how many are nickels and how many are dimes?

SOLUTION:

8) A man has $5.75 in nickels, dimes, and quarters. He has 3 times as many nickels as dimes, and 7 more quarters than dimes. How many of each coin does he have?

SOLUTION:

Don't Forget the Resources in the Back of the Book

There are several useful resources in the back of this book. If you have not taken the time to look through them, do so. When necessary, put them to good use.

Here is a list of the useful resources that are available to you.

- **Glossary**
- **Important Formulas**
- **Important Symbols**
- **Multiplication Table**
- **Commonly Used Prime Numbers**
- **Squares and Square Roots**
- **Fraction/Decimal Equivalents**
- **Solutions**

Chapter 8: Investment Problems

Everyone invests money in something. Some of the common financial investments are stocks, bonds, and savings accounts. It is important to know the following formula.

Principal x Rate = Income

Principal represents the amount of money invested. **Rate** represents the annual rate of interest, which is a percent. **Income** represents the annual income, which is sometimes called interest, earnings, or return. It is helpful to use a chart when solving investment problems. Also, using a calculator for computation is helpful. You might want to review the lessons on percent from chapter one. Remember the following when solving investment problems.

Helpful Hints:

- **First**, read the problem **carefully** and be sure that you fully understand it. Be sure you understand that which is **given**, and what is to be **found**.

- Assign a variable to one of the unknowns.

- Use a chart.

- Eliminate decimals from equations when possible. Multiplying each side of the equal sign by a power of ten does this.

- Use a calculator when completing the computation.

- Check your answers.

Chapter 8: Investment Problems

Solve each of the following.

1) **Jean invested $20,000, part at 7% and the rest at 5%. The interest earned in one year was $1,280. How much did Jean invest at each rate?**

Let x = the number of dollars invested at 7%

Investment	Principal x	Rate =	Income
7% Investment	x	.07	.07x
5% Investment	$20,000 − x	.05	.05($20,000 − x)

The total annual income = $1,280

$.07x + .05(\$20,000 - x) = \$1,280$ *Write the equation.*

$7x + 5(\$20,000 - x) = \$128,000$ *Mulitply both sides of the equal sign by 100 to eliminate the decimals.*

$7x + \$100,000 - 5x = \$128,000$

$7x - 5x + \$100,000 - \$100,000 = \$128,000 - \$100,000$ *Subtract $100,000 from each side*

$2x = \$28,000$

$x = \$14,000$

$x = \$14,000 @ 7\%$

$\$20,000 - x = \$6,000 @ 5\%$

$14,000 @ 7\%
$6,000 @ 5\%
Check your answers.

Chapter 8: Investment Problems

EXAMPLES

2) Robin made an investment. Part was invested at an interest rate of 6% and the remaining $2,000 was invested at 5%. The income for one year was $460. How much was invested at 6%?

Let x = the number of dollars invested at 6%

Investment	Principal x	Rate =	Income
6% Investment	x	.06	.06x
5% Investment	$2,000	.05	$100

The total annual income = $460

$$.06x + \$100 = \$460 \qquad \textit{Write the equation}$$

$$6x + \$10,000 = \$46,000 \qquad \textit{Mulitply both sides by 100}$$

$$6x = \$36,000$$

$$x = \$6,000$$

$$x = \$6,000 \ @ \ 6\%$$

$6,000 @ 6%
Check your answer

Notes:

_____

_____

_____

_____

_____

_____

_____

_____

Solve each of the following.

1) Donald invested some money at 6%. He also invested twice that amount at 5%. If the total annual income was $640, how much did he invest at each rate?

SOLUTION:

2) Susan invested $2,000. Part of it at 6% and the rest at 9%. If the annual income was $138, how much did she invest at each rate?

SOLUTION:

Chapter 8: Investment Problems

3) Sherry invested a sum of money at 6% and invested a second sum that was $1,500 greater than the first sum, at 5%. If the total annual income was $570, how much was invested at each rate?

SOLUTION:

4) A man invested $200,000. Part at 6% and the rest at 7%. The annual income was $13,400. How much was invested at each rate?

SOLUTION:

EXERCISES

5) Sheila invested a certain sum of money at 6%. She also invested a second sum, which was $2,000 less than the first sum, at 5%. Her total annual income was $890. How much did she invest at each rate?

SOLUTION:

6) Sam invested a sum of money at 6%. He invested a second sum, $1,000 more than the first sum at 7%. If the annual income was $1,110, how much did he invest at 6%? How much at 7%?

SOLUTION:

Chapter 8: Investment Problems

EXERCISES

7) Mrs. Jones invested $8,000. Part of it was invested at 3% and the rest at 4% annual interest. If the yearly return on these two investments is $275, how much did she invest at each rate?

SOLUTION:

8) A man invested $6,000. Some at 8%, and the rest at 10%. After one year, the total interest earned was $560. How much did he invest at each rate?

SOLUTION:

Algebra Word Problems Made Simple!

Are You Getting the Most Possible From This Book?

Here are a list of questions that you should ask yourself to determine whether you are taking the actions necessary to get the most possible out of using this book.

- Are you viewing each Online Tutorial Video lesson, and working along with the instructor?

- Are you carefully reading the Introduction to each lesson?

- Are you carefully reading the Helpful Hints section for each lesson?

- Are you neatly copying each Example and carefully writing down all the steps? Doing this will help you to more effectively understand the problem, and you will be more prepared for the Exercises.

- Are you neatly and carefully completing all of the Exercises, showing the work for each problem?

- Are you using the Solutions section to correct your work?

- Are you re-working the problems that you worked incorrectly, to find out what caused the mistake?

- Are you occasionally reviewing problems that you have completed?

- Are you using the Glossary and other resources located in the back of the book? Reading through the Glossary is a good way to review terms and their definitions.

Every question that you answered with yes represents an important step toward effectively learning algebra. It is up to you to use this book in a way that will benefit you the most!

Chapter 9: Proportions

A **ratio** compares two numbers or groups of objects. Notice in the diagram below that for every three circles there are four squares.

The ratio can be written in following ways:
3 to 4, **3:4**, and $\frac{3}{4}$. Each of these is read as "three to four".

In algebra, ratios are often written in fraction form. Remember that ratios expressed as fractions can be reduced to lowest terms.

Two equal ratios can be written as a **proportion**. The following is an example of a proportion: $\frac{4}{6} = \frac{2}{3}$

In a proportion, the **cross products** are equal.

For example, to determine whether $\frac{3}{4} = \frac{5}{6}$ is a proportion, simply **cross multiply**. $\frac{3}{4} \diagdown\!\!\!\!\diagup \frac{5}{6}$

$3 \times 6 = 18$, $4 \times 5 = 20$, $18 \neq 20$ It **is not** a proportion.

To determine whether $\frac{6}{9} = \frac{8}{12}$ is a proportion, again we can cross multiply.

$$\frac{6}{9} \diagdown\!\!\!\!\diagup \frac{8}{12}$$

$6 \times 12 = 72$, $9 \times 8 = 72$, $72 = 72$ It **is** a proportion.

It is easy to find the **missing number** in a proportion. For example, in the proportion $\frac{4}{n} = \frac{2}{3}$ we can easily find the value of n.

$\frac{4}{n} \diagdown\!\!\!\!\diagup \frac{2}{3}$

First, cross multiply: $2 \times n = 4 \times 3$

$2 \times n = 12$

Next, divide 12 by 2: $2\,\overline{\smash{\big)}\,12}^{\;6}$

$\boxed{n = 6}$

Chapter 9: Proportions

Proportions can often be used to solve word problems. All you need to do is set up a proportion and find the missing number. It is important to be consistent when you set the proportion up. For example, if you indicate hours at the top of one ratio, be sure that hours is on the top of the other ratio. Just remember to follow the following steps.

Helpful Hints:

- **First**, set up a proportion. Label the missing part with a variable (letter).

- **Second**, cross multiply to find the answer. In a proportion, the cross products are equal.

- Remember to be consistent. For example, if hours is on the bottom of one ratio, it has to be on the bottom of the other.

- Label your answer with a word or a short phrase.

EXAMPLE

Ratios and **proportions** can be used to solve problems.

Example: **A car can travel 384 km in six hours. How far can the car travel in eight hours?**

First set up a proportion. $\dfrac{384 \text{ km}}{6 \text{ hours}} = \dfrac{n \text{ km}}{8 \text{ hours}}$

Next, cross multiply. $6 \times n = 8 \times 384$
$6 \times n = 3{,}072$

Next, divide by six. $6 \overline{\smash{)}3072}$ gives 512 — $\boxed{n = 512}$

The car can travel 512 km in eight hours.

Chapter 9: Proportions

Use a proportion to solve each problem.

1) A car can travel 100 km on five liters of gas. How many liters will be needed to travel 40 km?

SOLUTION:

2) Two kg of chicken cost $7. How much will five kg cost?

SOLUTION:

EXERCISES

3) In a class, the ratio of boys to girls is four to three. If there are 20 boys in the class, how many girls are there?

SOLUTION:

4) A hiker takes three hours to go 24 km. At this rate, how far could he hike in five hours?

SOLUTION:

Chapter 9: Proportions

EXERCISES

5) Seven kg of nuts cost $5. How many kg of nuts can you buy with $2?

SOLUTION:

6) If 3 apples cost $1.19, how much would 18 apples cost?

SOLUTION:

Chapter 9: Proportions

7) A man can travel 13 km on his bicycle in 2 hours. At this rate, how far can he travel in 5 hours?

SOLUTION:

8) On a map, 1 cm = 30 km. How many cm on the map would represent 135 km?

SOLUTION:

Chapter 9: Proportions

9) A man earned $180 for working 14 hours. How many hours must he work to earn $300?

SOLUTION:

10) In a class, the ratio of boys to girls is 8 to 5. If there are 75 girls in the class, how many boys are there in the class?

SOLUTION:

EXERCISES

11) The weight of 80 meters of wire is 5 kg. What is the weight of 360 meters of wire?

SOLUTION:

12) The owner of a house worth $72,000 pays $2,000 in taxes. At this rate, how much will the taxes be on a house worth $54,000?

SOLUTION:

EXERCISES

13) A rectangular photo measures 15 cm by 20cm. It is enlarged so that the length of the shorter side is 21 cm. What would be the length of the longer side?

SOLUTION:

14) If 12 oranges cost $4.80, what would be the cost of 5 oranges?

SOLUTION:

Chapter 9: Proportions

EXERCISES

15) A man drove is car 400 miles in 8 hours. If he drives 700 miles the next day at the same rate, how long will it take to drive the 700 miles?

SOLUTION:

16) If four pounds of beef costs $4.80, how much will seven pounds cost?

SOLUTION:

Chapter 9: Proportions

EXERCISES

17) If six pounds of nuts cost $18, how many pounds of nuts can you buy with $12.?

SOLUTION:

18) On a map 3/4 of an inch represents 20 miles. What is the actual distance represented by 4 1/8 inches?

SOLUTION:

EXERCISES

19) The ratio of red marbles to blue marbles is five to two. If there are 15 red marbles, how many blue marbles are there?

SOLUTION:

20) A man who is 5 feet tall casts a shadow that is 8 feet long. She is standing next to a tree that is casting a shadow 72 feet long. How tall is the tree?

SOLUTION:

Chapter 10: Percents

Percent means "**per hundred**" or "**hundredths**". The symbol for percent is **%**. Percents can be expressed as **decimals** and **fractions**. The fraction form may sometime be reduced to its lowest terms. You will find that percent is used a lot in everyday life. The following examples show the close relationship between percents, fractions, and decimals.

$$25\% = .25 = \frac{25}{100} = \frac{1}{4} \qquad 8\% = .08 = \frac{8}{100} = \frac{2}{25}$$

When working with percents you may work with fractions or decimals. Use the one that is the easiest.

What good is math if you can't put it to good use? Percents are used often in a variety of everyday situations. Remember that there are basically three types of percent problems, and all you need to be able to do is identify them. Remember the following tips.

Helpful Hints:

- **Type I** problems ask you to find the **percent of a number**.

- **Type II** problems ask you to find the **percent**. In other words, there will be a percent sign (%) in the answer.

- **Type III** problems ask you to find the **whole** when the part and the percent are known.

Notes:

_____

_____

_____

_____

_____

Chapter 10: Percents

EXAMPLES

Type I

A man earns \$300 and spends 40% of it. How much does he spend?
(You know the whole.)

Find 40% of 300.

$$\begin{array}{r} \$300 \\ \times\ .4 \\ \hline \$120 \end{array}$$

He spends \$120.

Type II

In a class of 25 students, 15 of them are girls. What percent are girls?
(Your answer will have a percent sign—%.)

15 = what % of 25?

$$\frac{15}{25} = \frac{3}{5}$$

$$5\overline{)\,3.00} = .60 = 60\%$$

60% of the class are girls.

Type III

Five students got A's on a test. This is 20% of the class. How many students are there in the class? (You are finding the whole.)

5 = 20% of the class

5 ÷ .2

$$.2\overline{)\,5.0} = 25.$$

There are 25 students in the class.

Chapter 10: Percents

1) On a test with 25 questions, Al got 80% correct. How many questions did he get correct?

SOLUTION:

2) A player took 12 shots and made 9. What percent did he make?

SOLUTION:

Chapter 10: Percents

EXERCISES

3) A girl spent $5. This was 20% of her earnings. How much were her earnings?

SOLUTION:

4) Buying a $8,000 car requires a 20% down payment. How much is the down payment?

SOLUTION:

EXERCISES

5) 3 = 10% of what?

SOLUTION:

6) A team played 20 games and won 18. What % did they win?

SOLUTION:

EXERCISES

7) A farmer sold 50 cows. If this was 20% of his herd, how many cows were in his herd?

SOLUTION:

8) 20 = 80% of what?

SOLUTION:

9) Paul wants a bike that costs $400. If he has saved 60% of this amount, how much has he saved?

SOLUTION:

10) There are 400 students in a school. Fifty-five percent are girls. How many boys are there?

SOLUTION:

Chapter 10: Percents

EXERCISES

11) 12 is what % of 60?

SOLUTION:

12) Kelly earned 300 dollars and put 70% of it into the bank. How much did she put into the bank?

SOLUTION:

EXERCISES

13) An airline flight with 120 seats was filled to 65% of capacity. How many of the seats were occupied?

SOLUTION:

14) A salesman earned a commission of 15% for selling a car. If the commission was $1860, what was the price of the car?

SOLUTION:

EXERCISES

15) A student answered 35 questions correctly on a test a test with 40 questions. What percent of the problems did he answer correctly?

SOLUTION:

16) A man bought a television for $450.00. If the sales tax was 6%, what was the total cost of the television?

SOLUTION:

Chapter 10: Percents

EXERCISES

17) A man has a monthly income of $4500 per month. If 35% of his income is used to pay rent, how much is his rent?

SOLUTION:

18) The regular price of a jacket was $75.00. If John paid $60.00, what percent of the regular price did he save?

SOLUTION:

Algebra Word Problems Made Simple!

Chapter 10: Percents

19) A family went out for a dinner which cost $63.80. They left a 15% tip. How much was the tip, and what was the total bill?

> SOLUTION:

20) A women put $680 into her savings account. If this was 25% of her earnings for the month, what were her earnings for the month?

> SOLUTION:

Final Review

FINAL REVIEW

Solve each of the following.

1) Three times a certain number less 17 is equal to 28. Find the number.

SOLUTION:

2) One number is four times another. The difference between the two numbers is 48. Find the two numbers.

SOLUTION:

Final Review

FINAL REVIEW

3) If 12 more than 9 time a number is equal to 10 times that number plus 4, find the number.

SOLUTION:

4) The sum of two numbers is 88. One number is 24 greater than the other. Find the two numbers.

SOLUTION:

FINAL REVIEW

5) Two cars are 700 km apart. They drive towards each other. One travels at 40 km/hr and the other at 30 km/hr. How many hours before they meet?

SOLUTION:

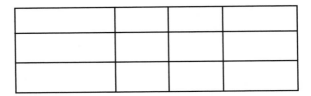

6) Two trains start from the same location and travel in opposite directions. One train travels 60 km/hr faster than the other. After 5 hours they are 1,500 km apart. Find the rate of each train.

SOLUTION:

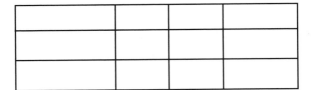

Final Review

FINAL REVIEW

7) Tom leaves his house traveling at 45 km/hr. Two hours later, his brother, Steve, leaves the house traveling 60 km/hr. How many hours will it take Steve to catch up with Tom?

SOLUTION:

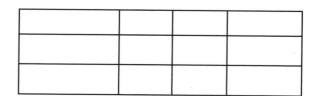

8) A man drove from his house to the lake at a rate of 60 km/hr. He returned home traveling at a rate of 45 km/hr. The entire trip took 7 hours. How far is it from his house to the lake?

SOLUTION:

Final Review

FINAL REVIEW

9) A store owner has some coffee worth $1.80 per kg and some worth $2.80 per kg. He wants to make 60 kg of a mixture that will be worth $2.20 per kg. How many kgs of each type of coffee should he use?

SOLUTION:

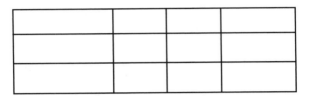

10) Steve can mow the lawn in 30 minutes by himself. Cathy can mow the same lawn in 50 minutes by herself. How long will it take to mow the lawn if they work together?

SOLUTION:

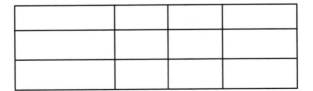

　　　　　　　　　　　　　　　　　　Algebra Word Problems Made Simple!

FINAL REVIEW

11) Peter and Paul can paint a room in 6 hours if they work together. If Peter works alone, it takes him 10 hours. How many hours will it take Paul to paint the room if he works alone?

SOLUTION:

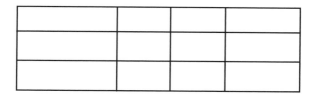

12) Steve has a collection of nickels and dimes. He has 8 more nickels than dimes. The value the coins is $2.80. How many coins of each type does he have?

SOLUTION:

FINAL REVIEW

13) Ellen has 40 coins which are all nickels and dimes. If the value of the coins is $3.20, how many of each coin does she have?

SOLUTION:

14) Maria is 5 years older than Sophia. Five years ago, Maria was two times as old as Sophia. What are their ages now?

SOLUTION:

FINAL REVIEW

15) Sonya is three times as old as Harry. In 18 years from now, Sonya will be twice as old as Harry. What are their present ages?

SOLUTION:

16) A man invested $9,000. Some was invested at 6% and some at 7%. The earnings for one year was $575. How much was invested at each rate?

SOLUTION:

Glossary

Abscissa The first number in an ordered pair that is assigned to a point on a coordinate plane. Also called the x-coordinate.

Absolute value The distance between 0 and a number on the number line. The absolute value of n is written $|n|$.

Algebra The branch of mathematics that uses letters and numbers to show the relationships between quantities.

Algebraic expression A mathematical expression which contains at least one variable. $2x$, $7x + 9$, $4ab$, and $\frac{x + 5}{2}$ are all algebraic expressions.

Associative properties
Addition: $(a + b) + c = a + (b + c)$
Multiplication: $(ab)c = a(bc)$

Axiom A property assumed to be true without proof. Also called a postulate.

Axis of symmetry A line that divides a parabola into two matching parts.

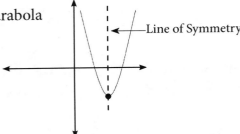

Base The number being multiplied. In an expression such as 4^2, 4 is the base.

Binomial A polynomial with two terms. $2x + 3y$ and $3x - 2y$ are binomials.

Cartesian coordinate system A system of graphing ordered pairs on a coordinate plane.

Coefficient A number that multiplies the variable. In the term 7x, 7 is the coefficient of x.

Commutative properties
Addition: $a + b = b + a$
Multiplication: $ab = ba$

Glossary

Completing the square A method for changing a quadratic expression into a perfect square trinomial.

For a quadratic expression in the form $x^2 + bx = c$, follow these steps:
1) Take half of b which is the coefficient of x
2) Square it.
3) Add the result to both sides of the equal sign.
4) The result will be a perfect square trinomial.

Compound inequality Two or more inequalities that are combined using the word "and" or the word "or."

Constant Specific numbers that do not change.

Coordinates An ordered pair of real numbers, which correspond to a point on a coordinate plane.

Coordinate plane The plane which contains the x and y-axes. It is divided into four quadrants.

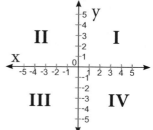

Degree of a polynomial The highest power of a variable that appears in a polynomial expression. The degree of $6x^3 - 4x + 1$ is 3.

Difference of two squares In the difference of two squares, $a^2 - b^2 = (a + b)(a - b)$.

Discriminant The value of $b^2 - 4ac$ is called the discriminant of the quadratic equation $ax + bx + c = 0$. It allows you to determine which quadratic equations have solutions and which ones do not.

Disjoint sets Sets which have no members in common.
(1, 2, 3) and (4, 5, 6) are disjoint sets.

Distance formula The distance D, between the points (x_1, y_1) and (x_2, y_2) on a coordinate plane is:
$$D = \sqrt{(x_2 - x_1)^2 + (y_2 - y_1)^2}$$

Glossary

Distributive Property For real numbers a, b, and c, $a(b + c) = ab + ac$.

Domain of a function The set of all first coordinates (x-values) of the ordered pairs that form the function.

Element of a set Member of a set.

Elimination Method A method of solving a system of linear equations using the following steps:
1) Put the variables on one side of the equal sign and the constant on the other, with the like terms lined up.
2) Add or subtract the equations to eliminate one of the variables. Sometimes it is necessary to multiply one of the equations by a constant first.
3) Solve the equation. Substitute the answer into either of the equations to get the value of the second variable.
4) Check by substituting the answers into the original equations.

Empty set The set that has no members. Also called the null set and is written $\emptyset$ or { }.

Equation A mathematical sentence that contains an equal sign (=) , and states that one expression is equal to another.

Equivalent expressions Expressions which represent the same number.

Evaluate an expression Finding the number an expression stands for by replacing each variable with its numerical value and the simplifying.

Exponent A number that indicates the number of times a given base is used as a factor. In the expression x^2, 2 is the exponent.

Extremes of a proportion In the proportion $\frac{a}{b} = \frac{c}{d}$, a and d are the extremes.

Factor A number or expression that is multiplied to get another number or expression. In the example $4 \times 3 = 12$, 4 and 3 are factors.

Formula An equation that states a relationship among quantities which are represented by variables. For example, the formula for the area of a rectangle is $A = l \times w$, where A = area, l = length, and w = width.

Function A set of ordered pairs which pairs each x-value with one and only one y-value. For example, F = {(0,2), (-1,6), (4,-2), (-3,4)} is a function.

Glossary

Function notation A way to describe a function that is defined by an equation.
In function notation the equation $y = 4x - 8$ is written as $f(x) = 4x - 8$, where $f(x)$ is read as "f of x" or "the value of f at x."

Graph To show the points named by numbers or ordered pairs on a number line or coordinate plane.

Greatest common factor The largest factor of two or more numbers or terms. Also written GCF.
The GCF of 15 and 10 is 5, since 5 is the largest number that divides evenly into both 10 and 15.
The GCF of 8ab and 6ab is 2ab.

Grouping symbols Symbols used to group mathematical expressions.
Examples include parentheses ( ), brackets [ ], braces { }, and fraction bars —.

Hypotenuse The side opposite the right angle in a right triangle.

Identity Properties
Addition: **a + 0 = 0 + a**
Multiplication: **1 x a = a**

Inequality A mathematical sentence that states that one expression is greater than or less than another. Inequality symbols are read as follows:
$<$ is less than, $\leq$ is less than or equal to, $>$ is greater than, $\geq$ is greater than or equal to

Integers Numbers in the set ...-3, -2, -1, 0, 1, 2, 3,... .

Intercept In the equation of a line, the y-intercept is the value of y when x is 0.
The x-intercept is the value of x when y is 0.

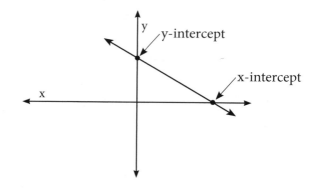

Glossary

Intersection of sets The intersection of two or more sets consists of the members included in all of them. A intersection B is written A∩B.
If set A = {1, 2, 3, 4} and set B = {1, 3, 5} then the intersection sets A and B would be the set {1, 3}.

Inverse operations Operations that "undo" each other. Addition and subtraction are inverse operations. Multiplication and division are inverse operations.

Irrational numbers A real number that cannot be written as the ratio of two integers. They are often represented by non-terminating, non-repeating decimals.
$\sqrt{2} = 1.4142135\ldots$ is an example of an irrational number.

Least Common Denominator (LCD) The least common multiple of the denominators of two or more fractions.
The LCD of $\frac{1}{8}$ and $\frac{1}{6}$ would be 24.
The LCD of $\frac{1}{6x^2}$ and $\frac{1}{4x}$ would be $12x^2$.

Least common multiple (LCM) The least common multiple of two or more expressions is the simplest expression that they will all divide into evenly. To do so, first find the LCM of the coefficients and then the highest degree of each variable and expression.
The LCM of $10x$ and $25x^2y$ would be $50x^2y$.
The LCM of $4(x-2)^2$ and $(x+2)^2(x-2)$ would be $4(x-2)^2(x+2)^2$.

Like Terms Terms that have the same variables raised to the same power.
$3xy^2$ and $9xy^2$ are like terms. The coefficients do not have to be the same.

Linear Equation An equation that can be written in the form **Ax + By = C**, where A and B are not both zero. The graph of a linear equation is a straight line.

Means of proportion In the proportion $\frac{a}{b} = \frac{c}{d}$, b and c are the means.

Midpoint formula The midpoint between the two points (x_1, y_1) and (x_2, y_2) is
$$M = \left(\frac{x_1 + x_2}{2}, \ \frac{y_1 + y_2}{2} \right)$$

Monomial A term that is a number, a variable, or the product of a number and one or more variables.
5, x, 4xy, 6xy are all examples of monomials.

Glossary

Multiple A multiple of a number is that number multiplied by an integer.
32 is a multiple of 4 since 4 x 8 = 32. Also, 4 and 8 are factors of 32.

Multiplicative inverse Two numbers whose product is one. They are also called reciprocals.
4 and $\frac{1}{4}$ and $-\frac{2}{3}$ and $-\frac{3}{2}$ are examples of multiplicative inverses.

Natural numbers Numbers in the set 1, 2, 3,... . Also called counting numbers.

Negative exponent For any non-zero number x, and any integer n,
$$x^{-n} = \frac{1}{x^n} \text{ and } \frac{1}{x^{-n}} = x^n$$

Negative number A number that is less than zero.
-5 and -3.45 are examples of negative numbers.

Negative slope When the graph of a line slopes down from left to right.

Null set The set that has no members. Also called the empty set which is written Ø and { }.

Number line A line that represents all real numbers with points.

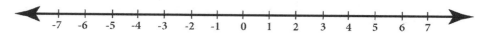

Open sentence A mathematical statement that contains at least one variable.
3x − 6 =12   n > 25   36 = x − 3 are all open sentences.

Ordered pair A pair of numbers (x, y) that represent a point on the coordinate plane. The first number is the x-coordinate and the second is the y-coordinate.

Order of operations The order of steps to be used when simplifying expressions.
1. Inside the grouping symbols.
2. Exponents
3. Multiply and divide in order from left to right.
4. Add and subtract in order from left to right.

Ordinate The second coordinate of an ordered pair. Also called the y-coordinate.

Origin The point where the x-axis and the y-axis intersect in a coordinate plane. Written as (0, 0).

Glossary

Parabola The U-shaped curve that is the graph of a quadratic function.

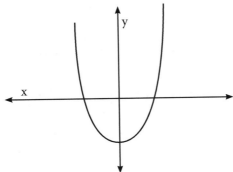

Parallel lines Lines in the same plane that do not intersect. Parallel lines have equal slopes.

Percent Part per hundred or hundredths. Written %.

Perfect square A number than can be expressed as the square of a rational number. The number 36 is a perfect square because it is the square of either 6 or -6.

Perfect square trinomial A trinomial that results from squaring a binomial. Written in the forms $a^2 + 2ab + b = (a + b)^2$ and $a^2 - 2ab + b = (a - b)^2$.

Perpendicular lines Lines in the same plane that intersect at a right (90°) angle. Perpendicular lines have slopes which are negative reciprocals of each other (the product of their slopes is -1).

Point-slope form An equation of a line in the form $y - y_1 = m(x - x_1)$ where m is the slope and (x_1, y_1) is a given point that lies on the line.

Polynomial An algebraic expression of one or more terms connected by plus (+) and minus (−) signs. A monomial has one term. A binomial has two terms. A trinomial has three terms.
3x is a monomial   3 + 5y is a binomial $x^2 + 4x + 3$ is a trinomial

Positive number A number that is greater than 0.
5 and 3.25 are examples of positive numbers.

Positive Slope When the graph of a line slopes up from left to right.

Postulate A property assumed to be true without proof. Also called an axiom

Power An expression that contains a base and an exponent. In the expression x^3, x is the base, and 3 is the exponent.

Glossary

Prime number A prime number is any whole number greater than 1, whose only factors are one and itself.

Prime factorization A whole number that is expressed as a product of its prime factors. The prime factorization of $100 = 2^2 \times 5^2$.

Proportion An equation that states that two ratios are equal.

Pythagorean theorem In a right triangle, if c is the hypotenuse and a and b are the other two legs, then $\mathbf{a^2 + b^2 = c^2}$.

Quadratic equation An equation that can be written in the form $\mathbf{ax^2 + bx + c = 0}$, where $a \neq 0$.

Quadratic formula The formula that can be used to solve any quadratic equation.

$$x = \frac{-b \pm \sqrt{b^2 - 4ac}}{2a}$$

Quadratic function A function that can be written in the form $\mathbf{y = ax^2 + bx + c}$, where $a \neq 0$.

Radical An expression that is written with a radical sign. The expressions $\sqrt{2}$, $\sqrt{x^2}$, and $\sqrt{25}$ are all radicals.

Radicand The expression that is inside the radical sign.

Range of a function The set of all second coordinates (y-values) of the ordered pairs that form the function.

Ratio A comparison of two numbers using division. Written a:b, a to b, and $\frac{a}{b}$.

Rational Expression A fraction whose numerator and denominator are polynomials. $\frac{x+3}{x-2}$ $\frac{x^2-6x+9}{x-3}$ $\frac{32xy^2}{8xy^2}$ are examples.

Rational numbers A number that can be expressed as the quotient of two integers. The denominator cannot equal zero.

Rationalizing the denominator Changing a fraction that has an irrational denominator to an equivalent fraction that has a rational denominator.

Glossary

Real numbers All positive and negative numbers and zero. This includes fractions and decimals.

Reciprocal The multiplicative inverse of a number. Their product is 1.
The reciprocal of 2 is $\frac{1}{2}$. The reciprocal of $-\frac{2}{3}$ is $-\frac{3}{2}$.

Relation Any set of ordered pairs.

Roots The solutions of a quadratic equation.

Rise The change in **y** going from one point to another on a coordinate plane.
The vertical change.

Run The change in **x** going from one point to another on a coordinate plane.
The horizontal change.

Scientific notation A number written as the product of a number between 1 and 10, and a power of ten. In scientific notation, $7,200 = 7.2 \times 10^3$.

Set A well-defined collection of objects.

Simplified expression The form of an expression with all like terms combined and written in its simplest form.

Slope The steepness of a line. The ratio of the rise (the change in the y direction) to the run (the change in the x direction).
For (x_1, y_1) and (x_2, y_2) which are any two points on a line, **slope** $= \frac{y_2 - y_1}{x_2 - x_1}$, $(x_2 \neq x_1)$

Slope-intercept form An equation of a line in the form **y = mx + b**. The slope is m.
The y-intercept is b.

Solution of an equation A number than can be substituted for the variable in an equation to make the equation true.

Solution of an equation containing two variables An ordered pair (x, y) that makes the equation a true statement.

Square Root If $a^2 = b$, then a is a square root of b. Square roots are written with a radical sign $\sqrt{}$.

Glossary

Standard form of a linear equation $Ax + By = C$, where A and B are not both zero.

Standard form of a quadratic equation $ax^2 + bx + c = 0$, where $a \neq 0$.

Subset If A and B are sets and all the members of set A are members of set B, then set A is a subset of set B.

Substitution method A method of solving a system of linear equations using the following steps:
1. Solve one of the equations for either x or y.
2. Substitute the expression from step 1 into the other equation and solve it for the other variable.
3. Take the value from step 2 and substitute it into either one of the original equations and solve it. You will now have two solutions.
4. Check the solutions in each of the original equations.

System of linear equations Two or more linear equations with the same variables.

System of linear inequalities Two or more linear inequalities in the same variable.

Terms Parts of an expression that are separated by addition or subtraction. Terms can be a number, a variable or a product or quotient of numbers and variables.

7, 3x, 4xy, and -3xy are all examples of terms.

Theorem A statement that can be proven to be true.

Transforming an equation To change an equation into an equivalent equation.

Trinomial A polynomial with three terms.
$x^2 + 2xy + y$ is an example of a trinomial.

Undefined rational expression A rational expression that has zero as a denominator. It is meaningless and is considered undefined.

Union of sets If A and B are sets, the union of A and B is the set whose members are found in set A, or set B, or both set A and set B. A union B is written $A \cup B$.

Variable A letter that represents a number

Variable expression Any expression that contains a variable.

Glossary

Venn diagram A type of diagram which shows how certain sets are related.

Vertex of a parabola The maximum or minimum point of a parabola

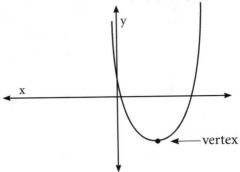

Vertical Line Test A way to tell whether a graph is a function. If a vertical line intersects a graph in more than one point, then the graph is **not** a function.

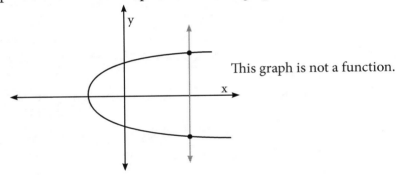

This graph is not a function.

Whole numbers Numbers in the set 0, 1, 2, 3,…

x-coordinate The first number in an ordered pair. Also called the abscissa.

x and y intercepts The points at which a graph intersects the x and y axes.

y-coordinate The second number in an ordered pair. Also called the ordinate.

Zero slope A **horizontal** line has zero slope. The slope of a **vertical** line is undefined.

Zero-product property If the product of two numbers is zero, then at least one of the numbers must equal zero.

Important Formulas

Algebraic Formulas	
Slope formula	$m = \frac{y_2 - y_1}{x_2 - x_1}$
Point-slope form	$y - y_1 = m(x - x_1)$
Slope-intercept form	$y = mx + b$
Standard form of a linear equation	$Ax + By = C$
Quadratic formula	$x = \frac{-b \pm \sqrt{b^2 - 4ac}}{2a}$
Pythagorean theorem	$a^2 + b^2 = c^2$
Distance formula	$D = \sqrt{(x_2 - x_1)^2 + (y_2 - y_1)^2}$
Midpoint formula	$M = (\frac{x_1 + x_2}{2}, \frac{y_1 + y_2}{2})$

Other Formulas	
Average speed	$r = \frac{d}{t}$
Interest	$I = p \times r \times t$

Geometric Formulas	
Perimeter of a polygon	P = the sum of the lengths of the sides
Circumference of a circle	$A = \Pi d$
Area of a rectangle	$A = lw$
Area of a square	$A = s^2$
Area of a parallelogram	$A = bh$
Area of a triangle	$A = \frac{1}{2}bh$
Area of a trapezoid	$A = \frac{1}{2}h(b_1 + b_2)$
Area of a circle	$A = \Pi r^2$
Volume of a cube	$V = s^3$
Volume of a rectangular prism	$V = lwh$

Important Symbols

$<$	less than	π	pi
$\leq$	less than or equal to	{ }	set
$>$	greater than	| |	absolute value
$\geq$	greater than or equal to	$.\overline{n}$	repeating decimal symbol
$=$	equal to	1/a	the reciprocal of a number
$\neq$	not equal to	%	percent
$\cong$	congruent to	(x,y)	ordered pair
$\approx$	is approximately equal to	f(x)	f of x, the value of f at x
( )	parenthesis	$\perp$	perpendicular
[ ]	brackets	| |	parallel to
{ }	braces	$\angle$	angle
...	and so on	$\in$	element of
• or ×	multiply	$\notin$	not an element of
∞	infinity	$\cap$	intersection
a^n	the nth power of a number	$\cup$	union
$\sqrt{}$	square root	$\subset$	subset of
Ø, { }	the empty set or null set	$\not\subset$	not a subset of
$\therefore$	therefore	$\triangle$	triangle
°	degree		

Algebra Word Problems Made Simple!

Multiplication Table

x	2	3	4	5	6	7	8	9	10	11	12
2	4	6	8	10	12	14	16	18	20	22	24
3	6	9	12	15	18	21	24	27	30	33	36
4	8	12	16	20	24	28	32	36	40	44	48
5	10	15	20	25	30	35	40	45	50	55	60
6	12	18	24	30	36	42	48	54	60	66	72
7	14	21	28	35	42	49	56	63	70	77	84
8	16	24	32	40	48	56	64	72	80	88	96
9	18	27	36	45	54	63	72	81	90	99	108
10	20	30	40	50	60	70	80	90	100	110	120
11	22	33	44	55	66	77	88	99	110	121	132
12	24	36	48	60	72	84	96	108	120	132	144

Commonly Used Prime Numbers

2	3	5	7	11	13	17	19	23	29
31	37	41	43	47	53	59	61	67	71
73	79	83	89	97	101	103	107	109	113
127	131	137	139	149	151	157	163	167	173
179	181	191	193	197	199	211	223	227	229
233	239	241	251	257	263	269	271	277	281
283	293	307	311	313	317	331	337	347	349
353	359	367	373	379	383	389	397	401	409
419	421	431	433	439	443	449	547	461	463
467	479	487	491	499	503	509	521	523	541
547	557	563	569	571	577	587	593	599	601
607	613	617	619	631	641	643	647	653	659
661	673	677	683	691	701	709	719	727	733
739	743	751	757	761	769	773	787	797	809
811	821	823	827	829	839	853	857	859	863
877	881	883	887	907	911	919	929	937	941
947	953	967	971	977	983	991	997	1009	1013

Squares and Square Roots

No.	Square	Square Root	No.	Square	Square Root	No.	Square	Square Root
1	1	1.000	51	2,601	7.141	101	10201	10.050
2	4	1.414	52	2,704	7.211	102	10,404	10.100
3	9	1.732	53	2,809	7.280	103	10,609	10.149
4	16	2.000	54	2,916	7.348	104	10,816	10.198
5	25	2.236	55	3,025	7.416	105	11,025	10.247
6	36	2.449	56	3,136	7.483	106	11,236	10.296
7	49	2.646	57	3,249	7.550	107	11,449	10.344
8	64	2.828	58	3,364	7.616	108	11,664	10.392
9	81	3.000	59	3,481	7.681	109	11,881	10.440
10	100	3.162	60	3,600	7.746	110	12,100	10.488
11	121	3.317	61	3,721	7.810	111	12,321	10.536
12	144	3.464	62	3,844	7.874	112	12,544	10.583
13	169	3.606	63	3,969	7.937	113	12,769	10.630
14	196	3.742	64	4,096	8.000	114	12,996	10.677
15	225	3.873	65	4,225	8.062	115	13,225	10.724
16	256	4.000	66	4,356	8.124	116	13,456	10.770
17	289	4.123	67	4,489	8.185	117	13,689	10.817
18	324	4.243	68	4,624	8.246	118	13,924	10.863
19	361	4.359	69	4,761	8.307	119	14,161	10.909
20	400	4.472	70	4,900	8.367	120	14,400	10.954
21	441	4.583	71	5,041	8.426	121	14,641	11.000
22	484	4.690	72	5,184	8.485	122	14,884	11.045
23	529	4.796	73	5,329	8.544	123	15,129	11.091
24	576	4.899	74	5,476	8.602	124	15,376	11.136
25	625	5.000	75	5,625	8.660	125	15,625	11.180
26	676	5.099	76	5,776	8.718	126	15,876	11.225
27	729	5.196	77	5,929	8.775	127	16,129	11.269
28	784	5.292	78	6,084	8.832	128	16,384	11.314
29	841	5.385	79	6,241	8.888	129	16,641	11.358
30	900	5.477	80	6,400	8.944	130	16,900	11.402
31	961	5.568	81	6,561	9.000	131	17,161	11.446
32	1,024	5.657	82	6,724	9.055	132	17,424	11.489
33	1,089	5.745	83	6,889	9.110	133	17,689	11.533
34	1,156	5.831	84	7,056	9.165	134	17,956	11.576
35	1,225	5.916	85	7,225	9.220	135	18,225	11.619
36	1,296	6.000	86	7,396	9.274	136	18,496	11.662
37	1,369	6.083	87	7,569	9.327	137	18,769	11.705
38	1,444	6.164	88	7,744	9.381	138	19,044	11.747
39	1,521	6.245	89	7,921	9.434	139	19,321	11.790
40	1,600	6.325	90	8,100	9.487	140	19,600	11.832
41	1,681	6.403	91	8,281	9.539	141	19,881	11.874
42	1,764	6.481	92	8,464	9.592	142	20,164	11.916
43	1,849	6.557	93	8,649	9.644	143	20,449	11.958
44	1,936	6.633	94	8,836	9.695	144	20,736	12.000
45	2,025	6.708	95	9,025	9.747	145	21,025	12.042
46	2,116	6.782	96	9,216	9.798	146	21,316	12.083
47	2,209	6.856	97	9,409	9.849	147	21,609	12.124
48	2,304	6.928	98	9,604	9.899	148	21,904	12.166
49	2,401	7.000	99	9,801	9.950	149	22,201	12.207
50	2,500	7.071	100	10,000	10.000	150	22,500	12.247

Algebra Word Problems Made Simple!

Fraction/Decimal Equivalents

Fraction	Decimal	Fraction	Decimal
$\frac{1}{2}$	0.5	$\frac{5}{10}$	0.5
$\frac{1}{3}$	0.3	$\frac{6}{10}$	0.6
$\frac{2}{3}$	0.6	$\frac{7}{10}$	0.7
$\frac{1}{4}$	0.25	$\frac{8}{10}$	0.8
$\frac{2}{4}$	0.5	$\frac{9}{10}$	0.9
$\frac{3}{4}$	0.75	$\frac{1}{16}$	0.0625
$\frac{1}{5}$	0.2	$\frac{2}{16}$	0.125
$\frac{2}{5}$	0.4	$\frac{3}{16}$	0.1875
$\frac{3}{5}$	0.6	$\frac{4}{16}$	0.25
$\frac{4}{5}$	0.8	$\frac{5}{16}$	0.3125
$\frac{1}{8}$	0.125	$\frac{6}{16}$	0.375
$\frac{2}{8}$	0.25	$\frac{7}{16}$	0.4375
$\frac{3}{8}$	0.375	$\frac{8}{16}$	0.5
$\frac{4}{8}$	0.5	$\frac{9}{16}$	0.5625
$\frac{5}{8}$	0.625	$\frac{10}{16}$	0.625
$\frac{6}{8}$	0.75	$\frac{11}{16}$	0.6875
$\frac{7}{8}$	0.875	$\frac{12}{16}$	0.75
$\frac{1}{10}$	0.1	$\frac{13}{16}$	0.8125
$\frac{2}{10}$	0.2	$\frac{14}{16}$	0.875
$\frac{3}{10}$	0.3	$\frac{15}{16}$	0.9375
$\frac{4}{10}$	0.4		

Solutions

Chapter 1
Page 12

Exercises

1) $2x - 7 = 12$

2) $3x + 2 = 30$

3) $2x + 5 = 14$

4) $4x - 6 = 10$

5) $4x - 5 = 12$

6) $\frac{1}{3}x - 4 = 2x + 8$

7) $2(x + 2) = 10$

8) $5x - 3 = 17$

9) $2x - 6 = 15$

10) $3x - 2 = 2x + 7$

11) $x + 4 = 7 + -12$

12) $\frac{x}{5} = 25$

13) $4x + 9 = 25$

14) $5x - 16 = x + 5$

15) $4x + 3 = 2x - 2$

16) $2(2y + 7) = 10$

17) $4(x + 9) = x - 8$

18) $\frac{x}{3} + 6 = 14$

19) $\frac{5x}{8} = 12$

20) $4x - 7 = 2x + 12$

Chapter 2
Page 21

Exercises

1) 8

2) 62

3) 5

4) 3

5) 45

6) 32, 64

7) John $18, Julie $54

8) 555 7th graders, 340 8th graders

9) Stan 22 candy bars, Stan 71 candy bars

10) 62, 63

11) 32, 33, 34

12) 51, 53, 55

13) 4

14) 9, 10, 11

15) 21, 84

16) 12, 24

17) 9, 22

18) 123, 135, 127

19) width 6, length 9

20) width 49, length 147

21) 12

Chapter 3
Page 32

Exercises

1) Joe = 7 km/hr, Ralph = 8 km/hr

2) 6 hours

3) Fast truck 97.5 km, Slow truck 82.5 km

4) 500 km/hr, 560 km/hr

5) 4 hours

6) 36 km

7) 6 hours

8) 8 km

9) 2 1/2 hours

10) Mr. Allen 28 mph, Mrs. Benson 20 mph

Chapter 4
Page 41

Exercises

1) 700 liters of 50 cent cleanser, 300 liters of 80 cent cleanser

2) 20 kg of vanilla, 40 kg of chocolate

3) 60 liters of 24% cream soup, 30 liters of 18% cream soup

4) 20 liters of 30% solution, 40 liters of 60% solution

5) 20 pounds

6) 42 ounces

7) 12 adults

8) 32 ounces

Algebra Word Problems Made Simple!

Solutions

Chapter 5
Page 49

Exercises

1) 2 hours
2) 12 minutes
3) 12 days
4) 6 hours
5) 12 hours
6) Dan works 6 days,
   Dave works 12 days
7) 3 hours
8) $7\frac{1}{5}$ hours

Chapter 6
Page 57

Exercises

1) Mother 24, Son 4
2) Zach 24, Moe 32
3) Chuck 12, Dave 24
4) Larry 15, Laura 40
5) Lance 17, Sam 24
6) Carly 30, Vica 60
7) Rhoda 6, Sophie 24
8) Ralph 8, Eric 32
9) Son 10, Father 50
10) Olivia 12, Jake 36

Chapter 7
Page 65

Exercises

1) 2 dimes, 8 quarters
2) 36 dimes, 24 quarters
3) 5 nickels, 35 dimes
4) 8 nickels, 24 dimes
5) 24 dimes, 16 quarters
6) 12 nickels, 7 dimes, 21 quarters
7) 6 nickels, 24 dimes
8) 24 nickels, 8 dimes, 15 quarters

Chapter 8
Page 73

Exercises

1) $4,000 at 6%,
   $8,000 at 5%
2) $1,400 at 6%,
   $600 at 9%
3) $4,500 at 6%,
   $6,000 at 5%
4) $60,000 at 6%,
   $140,000 at 7%
5) $7,000 at 5%,
   $9,000 at 6%
6) $8,000 at 6%,
   $9,000 at 7%
7) $4500 at 4%
   $3500 at 3%
8) $2,000 at 8%
   $4,000 at 10%

Solutions

Chapter 9
Page 80

Exercises

1) 2 liters
2) $17.50
3) 15 girls
4) 40 kilometers
5) $2\frac{4}{5}$ kilograms
6) $7.14
7) 32.5 km
8) 4.5 cm
9) $23\frac{1}{3}$ hrs = 23 hrs. 20 min.
10) 120 boys
11) 22.5 kg
12) $1500
13) 28 cm
14) $2.00
15) 14 hours
16) $8.40
17) 4 pounds)
18) 110 miles)
19) 6 blue marbles)
20) 45 feet)

Chapter 10
Page 92

Exercises

1) 20
2) 75%
3) $25
4) $1,600
5) 30
6) 90%
7) 250
8) 25
9) $240
10) 180
11) 20%
12) $210
13) 78 seats
14) $12,400
15) 87.5% or 87 1/2%
16) $477
17) $1575
18) 20%
19) tip $9.57, total $73.37
20) $2720

Final Review
Page 102

1) 15
2) 16 and 64
3) 8
4) 56 and 32
5) 10 hours
6) 120 km/hr   180 km/hr
7) 6 hours
8) 180 km
9) 36 kg or $1.80 coffee, 24 kg of $2.80 coffee
10) $18\frac{3}{4}$ minutes
11) 15 hours
12) 16 dimes, 24 nickels
13) 24 dimes, 16 nickels
14) Sophia 10, Maria 15
15) Sonya 54, Harry 18
16) $5,500 at 6%, $3,500 at 7%

OTHER MATH ESSENTIALS TITLES

Mastering Essential Math Skills: Book 1/Grades 4-5
ALSO AVAILABLE IN SPANISH LANGUAGE AND BILINGUAL EDITIONS

Mastering Essential Math Skills: Book 2/Middle Grades/High School
ALSO AVAILABLE IN SPANISH LANGUAGE AND BILINGUAL EDITIONS

Whole Numbers and Integers

Fractions

Decimals and Percents

Geometry

Problem Solving

Pre-Algebra Concepts
ALSO AVAILABLE IN SPANISH LANGUAGE AND BILINGUAL EDITIONS

No-Nonsense Algebra
ALSO AVAILABLE IN SPANISH LANGUAGE AND BILINGUAL EDITIONS

No-Nonsense Algebra Practice Workbook
ALSO AVAILABLE IN SPANISH LANGUAGE AND BILINGUAL EDITIONS

Math Refresher for Adults

Mastering Essential Math Skills: Graph Paper Notebooks
3 VERSIONS AVAILABLE FOR MATH, ALGEBRA, AND GEOMETRY

Try our free iphone app, Math Expert from Math Essentials

For more information go to www.mathessentials.net

Made in the USA
Monee, IL
19 January 2024

52025691R00072